THE VERTEX COVER ALGORITHM

ASHAY DHARWADKER

COPYRIGHT © 2006
INSTITUTE OF MATHEMATICS
H-501 PALAM VIHAR, GURGAON, HARYANA 122017, INDIA
www.dharwadker.org

ABSTRACT

We present a new polynomial-time algorithm for finding minimal vertex covers in graphs. It is shown that every graph with n vertices and maximum vertex degree Δ must have a minimum vertex cover of size at most $n - \lceil n/(\Delta+1) \rceil$ and that this condition is the best possible in terms of n and Δ. The algorithm finds a minimum vertex cover in all known examples of graphs. In view of the importance of the **P** versus **NP** question, we ask if there exists a graph for which the algorithm cannot find a minimum vertex cover. The algorithm is demonstrated by finding minimum vertex covers for several famous graphs, including two large benchmark graphs with hidden minimum vertex covers. We implement the algorithm in C++ and provide a demonstration program for Microsoft Windows.

The Demonstration Program

http://www.dharwadker.org/vertex_cover

3

CONTENTS

1. Introduction

In 1972, Karp [1] introduced a list of twenty-one **NP**-complete problems, one of which was the problem of finding a minimum vertex cover in a graph. Given a graph, one must find a smallest set of vertices such that every edge has at least one end vertex in the set. Such a set of vertices is called a minimum vertex cover of the graph and in general can be very difficult to find. For example, try to find a minimum vertex cover with seven vertices in the Frucht graph [2] shown below in Figure 1.1.

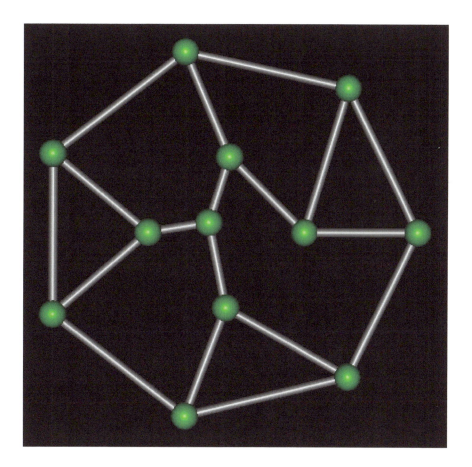

Figure 1.1. *Find a vertex cover with seven vertices*

We present a new polynomial-time **VERTEX COVER ALGORITHM** for finding minimal vertex covers in graphs. In Section 2, we provide precise **DEFINITIONS** of all the terminology used. In Section 3, we present a formal description of the **ALGORITHM** followed by a small example to show how the algorithm works step-by-step. In Section 4, we show that the algorithm has polynomial-time **COMPLEXITY**. In Section 5, we give a new condition of **SUFFICIENCY** for a graph to have a minimum vertex cover of a certain size. We prove that every graph with n vertices and maximum vertex degree Δ must have a minimum vertex cover of size at most $n - \lceil n/(\Delta+1) \rceil$ and that the algorithm will always find a vertex cover of at most this size. Furthermore, we prove that this condition is the

7

best possible in terms of n and Δ by explicitly constructing graphs for which the size of a minimum vertex cover is exactly $n-\lceil n/(\Delta+1)\rceil$. For all known examples of graphs, the algorithm finds a minimum vertex cover. In view of the importance of the **P** versus **NP** question **[3]**, we ask: *does there exist a graph for which this algorithm cannot find a minimum vertex cover?* In Section 6, we provide an **IMPLEMENTATION** of the algorithm as a C++ program, together with demonstration software for Microsoft Windows. In Section 7, we demonstrate the algorithm by finding minimum vertex covers for several **EXAMPLES** of famous graphs, including two large benchmark graphs with hidden minimum vertex covers. In Section 8, we list the **REFERENCES**.

2. Definitions

We begin with precise definitions of all the terminology and notation used in this presentation, following **[4]**. We use the usual notation $\lfloor x \rfloor$ to denote the *floor function* i.e. the greatest integer not greater than x and $\lceil x \rceil$ to denote the *ceiling function* i.e. the least integer not less than x.

A *simple graph G* with n vertices consists of a set of *vertices V*, with $|V| = n$, and a set of *edges E*, such that each edge is an unordered pair of distinct vertices. Note that the definition of G explicitly forbids *loops* (edges joining a vertex to itself) and *multiple edges* (many edges joining a pair of vertices), whence the set E must also be finite. We may *label* the vertices of G with the integers 1, 2, …, n. If the unordered pair of vertices $\{u, v\}$ is an edge in G, we say that u is a *neighbor* of v and write $uv \in E$. Neighborhood is clearly a symmetric relationship: $uv \in E$ if and only if $vu \in E$. The *degree* of a vertex v, denoted by $d(v)$, is the number of neighbors of v. The *maximum degree* over all vertices of G is denoted by Δ. The *adjacency matrix* of G is an $n \times n$ matrix with the entry in row u and column v equal to 1 if $uv \in E$ and equal to 0 otherwise. A *clique Q* of G is a set of vertices such that every unordered pair of vertices in Q is an edge. An *independent set S* of G is a set of vertices such that no unordered pair of vertices in S is an edge. A *vertex cover C* of G is a set of vertices such that for every edge $\{u,v\}$ of G at least one of u or v is in C. Given a vertex cover C of G and a vertex v in C, we say that v is *removable* if the set $C-\{v\}$ is still a vertex cover of G. Denote by $\rho(C)$ the *number of removable vertices* of a vertex cover C of G. A *minimal vertex cover* has no removable vertices. A *minimum vertex cover* is a vertex cover with the least number of vertices. Note that a minimum vertex cover is always minimal but not necessarily vice versa.

An *algorithm* is a problem-solving method suitable for implementation as a computer program. While designing algorithms we are typically faced with a number of different approaches. For small problems, it hardly matters which approach we use, as long as it is one that solves the problem correctly. However, there are many problems for which the only known algorithms take so long to compute the solution that they are practically useless. A *polynomial-time algorithm* is one whose number of computational steps is always bounded by a polynomial function of the size of the input. Thus, a polynomial-time algorithm is one that is actually useful in practice. The class of all such problems that have polynomial-time algorithms is denoted by **P**. For some problems, there are no known polynomial-time algorithms but these problems do have *nondeterministic polynomial-time algorithms*: try all candidates for solutions simultaneously and for each given candidate, verify whether it is a correct solution in polynomial-time. The class of all such problems is denoted by **NP**. Clearly **P** $\subseteq$ **NP**. On the other hand, there are problems that are known to be in **NP** and are such that any polynomial-time algorithm for them can be transformed (in polynomial-time) into a polynomial-time algorithm for every problem in **NP**. Such problems are called **NP**-*complete*. The problem of finding a minimum vertex cover is known to be **NP**-complete [1]. Thus, if we are able to show the existence of a polynomial-time algorithm that finds a minimum vertex cover in any graph, we could prove that **P** = **NP**. The present algorithm is, so far as we know, a promising candidate for the task. One of the greatest unresolved problems in mathematics and computer science today is whether **P** = **NP** or **P** $\neq$ **NP** [3].

3. Algorithm

We now present a formal description of the algorithm. This is followed by a small example illustrating the steps of the algorithm. We start by defining two procedures.

3.1. Procedure. Given a simple graph G with n vertices and a vertex cover C of G, if C has no removable vertices, output C. Else, for each removable vertex v of C, find the number $\rho(C-\{v\})$ of removable vertices of the vertex cover $C-\{v\}$. Let $v_{\max}$ denote a removable vertex such that $\rho(C-\{v_{\max}\})$ is a maximum and obtain the vertex cover $C-\{v_{\max}\}$. Repeat until the vertex cover has no removable vertices.

3.2. Procedure. Given a simple graph G with n vertices and a minimal vertex cover C of G, if there is no vertex v in C such that v has exactly one neighbor w outside C, output C. Else, find a vertex v in C such that v has exactly one neighbor w outside C. Define $C^{v,w}$ by removing v from C and adding w to C. Perform procedure 3.1 on $C^{v,w}$ and output the resulting vertex cover.

3.3. Algorithm. Given as input a simple graph G with n vertices labeled 1, 2, …, n, search for a vertex cover of size at most k. At each stage, if the vertex cover obtained has size at most k, then stop.

- **Part I.** For $i = 1, 2, ..., n$ in turn
  - Initialize the vertex cover $C_i = V-\{i\}$.
  - Perform procedure 3.1 on C_i.
  - For $r = 1, 2, ..., n-k$ perform procedure 3.2 repeated r times.
  - The result is a minimal vertex cover C_i.
- **Part II.** For each pair of minimal vertex covers C_i, C_j found in Part I
  - Initialize the vertex cover $C_{i,j} = C_i \cup C_j$.
  - Perform procedure 3.1 on $C_{i,j}$.

9

- o   For $r = 1, 2, ..., n−k$ perform procedure 3.2 repeated r times.
- o   The result is a minimal vertex cover $C_{i,j}$.

3.4. Example. We demonstrate the steps of the algorithm with a small example. The input is the Frucht graph [2] shown below with $n = 12$ vertices labled

$$V = \{1, 2, 3, 4, 5, 6, 7, 8, 9, 10, 11, 12\}.$$

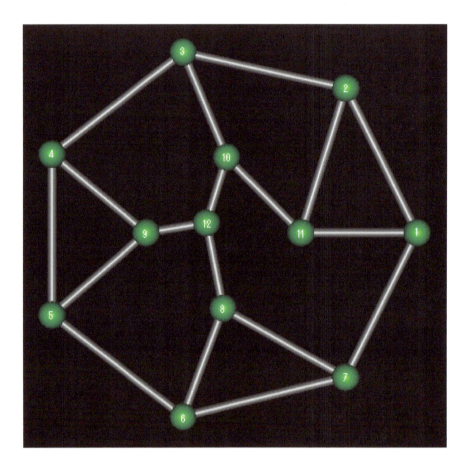

Figure 3.1. A small example to demonstrate the steps of the algorithm

We search for a vertex cover of size at most $k = 7$. Part I for $i = 1$ and $i = 2$ yields vertex covers C_1 and C_2 of size 8, so we give the details starting from $i = 3$. We initialize the vertex cover as

$$C_3 = V−\{3\} = \{1, 2, 4, 5, 6, 7, 8, 9, 10, 11, 12\}.$$

We now perform procedure 3.1. Here are the results in tabular form:

Vertex Cover C_3 = {1, 2, 4, 5, 6, 7, 8, 9, 10, 11, 12}. Size: 11.

Removable vertex v of C_3	Removable vertices of $C_3-\{v\}$	$\rho(C_3-\{v\})$
1	5, 6, 8, 9, 12	5
5	1, 7, 8, 11, 12	5
6	1, 9, 11, 12	4
7	5, 9, 11, 12	4
8	1, 5, 9, 11	4
9	1, 6, 7, 8, 11	5
11	5, 6, 7, 8, 9, 12	6
12	1, 5, 6, 7, 11	5

Maximum $\rho(C_3-\{v\})$ = 6 for v = 11. Remove vertex 11 from C_3.

Vertex Cover C_3 = {1, 2, 4, 5, 6, 7, 8, 9, 10, 12}. Size: 10.

Removable vertex v of C_3	Removable vertices of $C_3-\{v\}$	$\rho(C_3-\{v\})$
5	7, 8, 12	3
6	9, 12	2
7	5, 9, 12	3
8	5, 9	2
9	6, 7, 8	3
12	5, 6, 7	3

Maximum $\rho(C_3-\{v\})$ = 3 for v = 5. Remove vertex 5 from C_3.

Vertex Cover C_3 = {1, 2, 4, 6, 7, 8, 9, 10, 12}. Size: 9.

Removable vertex v of C_3	Removable vertices of $C_3-\{v\}$	$\rho(C_3-\{v\})$
7	12	1
8	None	0
12	7	1

Maximum $\rho(C_3 - \{v\}) = 1$ for $v = 7$. Remove vertex 7 from C_3.

Vertex Cover $C_3 = \{1, 2, 4, 6, 8, 9, 10, 12\}$. Size: 8.

Removable vertex v of C_3	Removable vertices of $C_3 - \{v\}$	$\rho(C_3 - \{v\})$
12	None	0

Maximum $\rho(C_3 - \{v\}) = 0$ for $v = 12$. Remove vertex 12 from C_3.

We obtain a minimal vertex cover

$$C_3 = \{1, 2, 4, 6, 8, 9, 10\}$$

of the requested size $k = 7$ and the algorithm terminates.

4. Complexity

We shall now show that the algorithm terminates in polynomial-time, by specifying a polynomial of the number of vertices n of the input graph, that is an upper bound on the total number of computational steps performed by the algorithm. Note that we consider

- checking whether a given pair of vertices is connected by an edge in G, and
- comparing whether a given integer is less than another given integer

to be *elementary computational steps*.

4.1. Proposition. Given a simple graph G with n vertices and a vertex cover C, procedure 3.1 takes at most n^5 steps.

Proof. Checking whether a particular vertex is removable takes at most n^2 steps, since the vertex has less than n neighbors and for each neighbor it takes less than n steps to check whether it is in the vertex cover. For a particular vertex cover, finding the number ρ of removable vertices takes at most $n^3 = nn^2$ steps, since for each of the at most n vertices in the vertex cover we must check whether it is removable or not. For a particular vertex cover, finding a vertex for which ρ is maximum then takes at most $n^4 = nn^3$ steps, since the vertex cover has at most n vertices. Procedure 3.1 terminates when at most n vertices are removed, so it takes a total of at most $n^5 = nn^4$ steps. $\square$

4.2. Proposition. Given a simple graph G with n vertices and a minimal vertex cover C, procedure 3.2 takes at most $n^5 + n^2 + 1$ steps.

Proof. To find a vertex v in C that has exactly one neighbor w outside C takes at most n^2 steps, since there are less than n vertices in C and we must find out if at least one of the less than n neighbors of any such vertex are outside C. If such a vertex v has been found, it takes one step to exchange v and w. Thereafter, by proposition 4.1, it takes at most n^5 steps to perform procedure 3.1 on the resulting vertex cover. Thus, procedure 3.2 takes at most n^2+1+n^5 steps. $\square$

4.3. Proposition. Given a simple graph G with n vertices, part I of the algorithm takes at most $n^7+n^6+n^4+n^2$ steps.

Proof. At each turn, procedure 3.1 takes at most n^5 steps by proposition 4.1. Then procedure 3.2 is performed at most n times, since k can be at most n. This, by proposition 4.2, takes at most $n(n^5+n^2+1) = n^6+n^3+n$ steps. So, at each turn, at most $n^5+n^6+n^3+n$ steps are executed. There are n turns for $i = 1, 2, ..., n$, so part I performs a total of at most $n(n^5+n^6+n^3+n) = n^6+n^7+n^4+n^2$ steps. $\square$

4.4. Proposition. Given a simple graph G with n vertices, the algorithm takes less than $n^8+2n^7+n^6+n^5+n^4+n^3+n^2$ steps to terminate.

Proof. There are less than n^2 distinct pairs of minimal vertex covers found by part I, that are treated in turn. Similar to the proof of proposition 4.3, part II takes less than $n^2(n^5+n^6+n^3+n) = n^7+n^8+n^5+n^3$. Hence, part I and part II together take less than a grand total of $(n^7+n^6+n^4+n^2)+(n^8+n^7+n^5+n^3) = n^8+2n^7+n^6+n^5+n^4+n^3+n^2$ steps to terminate. $\square$

4.5. Remark. These are pessimistic upper bounds for the worst possible cases. The actual number of steps taken by the algorithm to terminate will depend on both n and k. For larger values of k, the algorithm terminates much faster. In almost all of the examples in section 7, one or two steps of part I already find a minimum vertex cover. Only the second benchmark, Witzel's graph 7.20, requires part II of the algorithm to find a minimum vertex cover.

5. Sufficiency

The algorithm may be applied to any simple graph and will always terminate in polynomial-time, finding many minimal vertex covers. The propositions below establish sufficient conditions on the input graph which guarantee that the algorithm will find minimal vertex covers of a certain size. Specifically, we prove that every graph with n vertices and maximum vertex degree Δ must have a minimum vertex cover of size at most $n-\lceil n/(\Delta+1)\rceil$ and that the algorithm will always find a vertex cover of at most this size. Furthermore, we prove that this condition is the best possible in terms of n and Δ by explicitly constructing graphs for which the size of a minimum vertex cover is exactly $n-\lceil n/(\Delta+1)\rceil$. The proofs use two fundamental axioms: Euclid's Division Lemma [5] and the Pigeonhole Principle [6].

Euclid's Division Lemma. Given a positive integer m and any integer n, there exist unique integers q and r with $0 \leq r < m$ such that $n = qm + r$.

Pigeonhole Principle. If l letters are distributed into p pigeonholes, then some pigeonhole receives at least $\lceil l/p \rceil$ letters and some pigeonhole receives at most $\lfloor l/p \rfloor$ letters.

5.1. Proposition. Given a simple graph G with n vertices and an initial vertex cover C. At each stage of procedure 3.1, if the vertex cover C has l vertices and the maximum degree among the vertices outside C is less than $\lceil l/(n-l) \rceil$, then procedure 3.1 produces a strictly smaller vertex cover.

Proof. By contradiction. Suppose the vertex cover C is minimal. Then there are no removable vertices and every vertex in C must have a neighbor outside C. Thus there are at least l edges (letters) with one end vertex in C and the other end vertex outside C, there being exactly $p = n-l$ vertices outside C (pigeonholes). By the pigeonhole principle, some vertex outside C must receive at least $\lceil l/p \rceil$ edges contradicting the hypothesis that the maximum degree among the vertices outside C is less than $\lceil l/p \rceil$. $\square$

5.2. Proposition. Given a vertex cover C of G, procedure 3.1 always produces a minimal vertex cover of G.

Proof. Procedure 3.1 terminates only when there are no removable vertices. By definition, the resulting vertex cover must be minimal. $\square$

5.3. Proposition. Given a simple graph G with n vertices and an initial minimal vertex cover C. If the minimal vertex cover C has m vertices and the maximum degree among the vertices outside C is less than $\lceil 2m/(n-m) \rceil$, then there exists a vertex v in C such that v has exactly one neighbor w outside C and procedure 3.2 produces a minimal vertex cover different from C and of size less than or equal to the size of C.

Proof. By contradiction. Note that since C is minimal, there are no removable vertices and every vertex in C has at least one neighbor outside C. Suppose every vertex in C has more than one neighbor outside C. Then there are at least $l = 2m$ edges (letters) with one end vertex in C and the other end vertex outside C, there being exactly $p = n-m$ vertices outside C (pigeonholes). By the pigeonhole principle, some vertex outside C must receive at least $\lceil l/p \rceil$ edges contradicting the hypothesis that the maximum degree among the vertices outside C is less than $\lceil l/p \rceil$. Thus, there exists a vertex v in C such that v has exactly one neighbor w outside C. Now since procedure 3.2 exchanges v and w, a vertex cover different from C but of the same size as C is created. Note that in the process some vertices of the vertex cover might have become removable. Then, procedure 3.2 applies procedure 3.1 that produces a minimal vertex cover different from C and of size less than or equal to the size of C. $\square$

5.4. Proposition. Given a simple graph G with n vertices and maximum vertex degree Δ, the algorithm always finds a minimal vertex cover of size at most $n - \lceil n/(\Delta+1) \rceil$.

14

Proof. Consider any one turn of part I in the algorithm. After t vertices have been removed from a total of n, the vertex cover C has $l = n-t$ vertices and the maximum degree among the vertices outside C is certainly less than or equal to Δ. By proposition 5.1, if Δ is less than $\lceil l/(n-l) \rceil = \lceil (n-t)/(n-(n-t)) \rceil = \lceil (n-t)/t \rceil = \lceil (n/t)-1 \rceil$, then a strictly smaller vertex cover is produced by the removal of a vertex. Hence, as long as t is less than $\lceil n/(\Delta+1) \rceil$, a vertex can still be removed and procedure 1 continues. Thus, at least $\lceil n/(\Delta+1) \rceil$ are removed, leaving a vertex cover of size at most $n-\lceil n/(\Delta+1) \rceil$. By propositions 5.1, 5.2 and 5.3, all of the vertex covers produced by the algorithm are minimal and of size at most $n-\lceil n/(\Delta+1) \rceil$. □

5.5. Proposition. A simple graph G with n vertices and maximum vertex degree Δ has a minimal vertex cover of size at most $n-\lceil n/(\Delta+1) \rceil$.

Proof. By proposition 5.4, the algorithm finds a minimal vertex cover of size at most $n-\lceil n/(\Delta+1) \rceil$. □

5.6. Proposition. Given any positive integers n and Δ such that $0 < \Delta < n$, there exists a graph G with maximum vertex degree Δ and a minimum vertex cover of size $n-\lceil n/(\Delta+1) \rceil$. For any such graph the algorithm always finds a minimum vertex cover.

Proof. Let $n = q(\Delta+1)+r$ with $0 \le r < \Delta+1$ by Euclid's division lemma. There are two cases.

- *Case* 1. Suppose $r = 0$. Define G to be the graph consisting of q disjoint cliques $Q_1, ..., Q_q$ with $\Delta+1$ vertices each. Then G is a graph with maximum vertex degree Δ. Suppose $C_{minimum}$ is a minimum vertex cover of G. Then $C_{minimum}$ must contain all but one vertex from each clique, i.e. the size of $C_{minimum}$ must be at least $q\Delta$. On the other hand, by proposition 5.4, the algorithm finds a minimal vertex cover C of size at most $n-\lceil n/(\Delta+1) \rceil = q(\Delta+1)-\lceil q(\Delta+1)/(\Delta+1) \rceil = q(\Delta+1)-q = q\Delta$. Thus, the size of C and $C_{minimum}$ must be the same, i.e. $n-\lceil n/(\Delta+1) \rceil$.
- *Case* 2. Suppose r is positive. Define G to be the graph consisting of q disjoint cliques $Q_1, ...,Q_q$ with $\Delta+1$ vertices each and a disjoint clique R with r vertices. Then G is a graph with maximum vertex degree Δ. Suppose $C_{minimum}$ is a minimum vertex cover of G. Then $C_{minimum}$ must contain all but one vertex from each clique, i.e. the size of $C_{minimum}$ must be at least $q\Delta+r-1$. On the other hand, by proposition 5.4, the algorithm finds a minimal vertex cover C of size at most $n-\lceil n/(\Delta+1) \rceil = q(\Delta+1)+r-\lceil (q(\Delta+1)+r)/(\Delta+1) \rceil = q(\Delta+1)+r-q-\lceil r/(\Delta+1) \rceil = q\Delta+q+r-q-1 = q\Delta+r-1$, using the fact that $\lceil r/(\Delta+1) \rceil = 1$ since $0 < r < \Delta+1$. Thus, the size of C and $C_{minimum}$ must be the same, i.e. $n-\lceil n/(\Delta+1) \rceil$. □

5.7. Question. For all known examples of graphs, the algorithm finds a minimum vertex cover. In view of the importance of the **P** versus **NP** question [3], we ask: *does there exist a graph for which this algorithm cannot find a minimum vertex cover?*

6. Implementation

We demonstrate the algorithm with a C++ program following the style of **[7]**. The demonstration program package **[download]** contains a detailed help file and section 7 gives several examples of input/output files for the program.

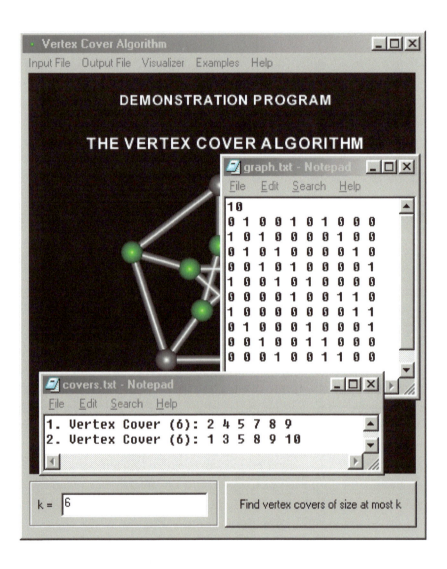

Figure 6.1. *Demonstration program for Microsoft Windows* **[download]**

```cpp
vertex_cover.cpp
#include <iostream>
#include <fstream>
#include <string>
#include <vector>
using namespace std;

bool removable(vector<int> neighbor, vector<int> cover);
int max_removable(vector<vector<int> > neighbors, vector<int> cover);
vector<int> procedure_1(vector<vector<int> > neighbors, vector<int>
cover);
vector<int> procedure_2(vector<vector<int> > neighbors, vector<int>
cover, int k);
int cover_size(vector<int> cover);
ifstream infile ("graph.txt");
ofstream outfile ("covers.txt");

int main()
{
 //Read Graph
 cout<<"Vertex Cover Algorithm."<<endl;
 int n, i, j, k, p, q, r, s, min, edge, counter=0;
 infile>>n;
 vector< vector<int> > graph;
 for(i=0; i<n; i++)
 {
  vector<int> row;
  for(j=0; j<n; j++)
  {
   infile>>edge;
   row.push_back(edge);
  }
  graph.push_back(row);
 }
 //Find Neighbors
 vector<vector<int> > neighbors;
 for(i=0; i<graph.size(); i++)
 {
  vector<int> neighbor;
  for(j=0; j<graph[i].size(); j++)
  if(graph[i][j]==1) neighbor.push_back(j);
  neighbors.push_back(neighbor);
 }
 cout<<"Graph has n = "<<n<<" vertices."<<endl;
 //Read minimum size of Vertex Cover wanted
 cout<<"Find a Vertex Cover of size at most k = ";
 cin>>k;
 //Find Vertex Covers
 bool found=false;
 cout<<"Finding Vertex Covers..."<<endl;
 min=n+1;
 vector<vector<int> > covers;
 vector<int> allcover;
 for(i=0; i<graph.size(); i++)
 allcover.push_back(1);
```

17

```cpp
for(i=0; i<allcover.size(); i++)
{
 if(found) break;
 counter++; cout<<counter<<". ";  outfile<<counter<<". ";
 vector<int> cover=allcover;
 cover[i]=0;
 cover=procedure_1(neighbors,cover);
 s=cover_size(cover);
 if(s<min) min=s;
 if(s<=k)
 {
  outfile<<"Vertex Cover ("<<s<<"): ";
  for(j=0; j<cover.size(); j++) if(cover[j]==1) outfile<<j+1<<" ";
  outfile<<endl;
  cout<<"Vertex Cover Size: "<<s<<endl;
  covers.push_back(cover);
  found=true;
  break;
 }
 for(j=0; j<n-k; j++)
 cover=procedure_2(neighbors,cover,j);
 s=cover_size(cover);
 if(s<min) min=s;
 outfile<<"Vertex Cover ("<<s<<"): ";
 for(j=0; j<cover.size(); j++) if(cover[j]==1) outfile<<j+1<<" ";
 outfile<<endl;
 cout<<"Vertex Cover Size: "<<s<<endl;
 covers.push_back(cover);
 if(s<=k){ found=true; break; }
}
//Pairwise Unions
 for(p=0; p<covers.size(); p++)
 {
  if(found) break;
  for(q=p+1; q<covers.size(); q++)
  {
   if(found) break;
   counter++; cout<<counter<<". ";  outfile<<counter<<". ";
   vector<int> cover=allcover;
   for(r=0; r<cover.size(); r++)
   if(covers[p][r]==0 && covers[q][r]==0) cover[r]=0;
   cover=procedure_1(neighbors,cover);
   s=cover_size(cover);
   if(s<min) min=s;
   if(s<=k)
   {
    outfile<<"Vertex Cover ("<<s<<"): ";
    for(j=0; j<cover.size(); j++) if(cover[j]==1) outfile<<j+1<<" ";
    outfile<<endl;
    cout<<"Vertex Cover Size: "<<s<<endl;
    found=true;
    break;
   }
   for(j=0; j<k; j++)
   cover=procedure_2(neighbors,cover,j);
   s=cover_size(cover);
   if(s<min) min=s;
```

```
    outfile<<"Vertex Cover ("<<s<<"): ";
    for(j=0; j<cover.size(); j++) if(cover[j]==1) outfile<<j+1<<" ";
    outfile<<endl;
    cout<<"Vertex Cover Size: "<<s<<endl;
    if(s<=k){ found=true; break; }
    }
  }
 if(found) cout<<"Found Vertex Cover of size at most "<<k<<"."<<endl;
 else cout<<"Could not find Vertex Cover of size at most
"<<k<<"."<<endl
<<"Minimum Vertex Cover size found is "<<min<<"."<<endl;
 cout<<"See cover.txt for results."<<endl;
 system("PAUSE");
 return 0;
}


bool removable(vector<int> neighbor, vector<int> cover)
{
 bool check=true;
 for(int i=0; i<neighbor.size(); i++)
 if(cover[neighbor[i]]==0)
 {
  check=false;
  break;
 }
 return check;
}


int max_removable(vector<vector<int> > neighbors, vector<int> cover)
{
 int r=-1, max=-1;
 for(int i=0; i<cover.size(); i++)
 {
  if(cover[i]==1 && removable(neighbors[i],cover)==true)
  {
   vector<int> temp_cover=cover;
   temp_cover[i]=0;
   int sum=0;
   for(int j=0; j<temp_cover.size(); j++)
   if(temp_cover[j]==1 && removable(neighbors[j], temp_cover)==true)
   sum++;
   if(sum>max)
   {
    max=sum;
    r=i;
   }
  }
 }
 return r;
}


vector<int> procedure_1(vector<vector<int> > neighbors, vector<int>
cover)
{
 vector<int> temp_cover=cover;
 int r=0;
```

```
 while(r!=-1)
 {
  r= max_removable(neighbors,temp_cover);
  if(r!=-1) temp_cover[r]=0;
 }
 return temp_cover;
}


vector<int> procedure_2(vector<vector<int> > neighbors, vector<int>
cover, int k)
{
 int count=0;
 vector<int> temp_cover=cover;
 int i=0;
 for(int i=0; i<temp_cover.size(); i++)
 {
  if(temp_cover[i]==1)
  {
   int sum=0, index;
   for(int j=0; j<neighbors[i].size(); j++)
   if(temp_cover[neighbors[i][j]]==0) {index=j; sum++;}
   if(sum==1 && cover[neighbors[i][index]]==0)
   {
    temp_cover[neighbors[i][index]]=1;
    temp_cover[i]=0;
    temp_cover=procedure_1(neighbors,temp_cover);
    count++;
   }
   if(count>k) break;
  }
 }
 return temp_cover;
}


int cover_size(vector<int> cover)
{
 int count=0;
 for(int i=0; i<cover.size(); i++)
 if(cover[i]==1) count++;
 return count;
}
```

Figure 6.2. C++ program for the vertex cover algorithm [**download**]

7. Examples

We demonstrate the algorithm by running the program on several famous graphs and two large benchmark graphs with hidden minimum vertex covers. In each case, the algorithm finds a minimum vertex cover in polynomial-time.

7.1. The Tetrahedron [8]. We run the program on the graph of the Tetrahedron with $n = 4$ vertices. The algorithm finds a minimum vertex cover of size $k = 3$.

```
graph.txt
4
0 1 1 1
1 0 1 1
1 1 0 1
1 1 1 0
```

```
cover.txt
Vertex Cover (3): 1 3 4
```

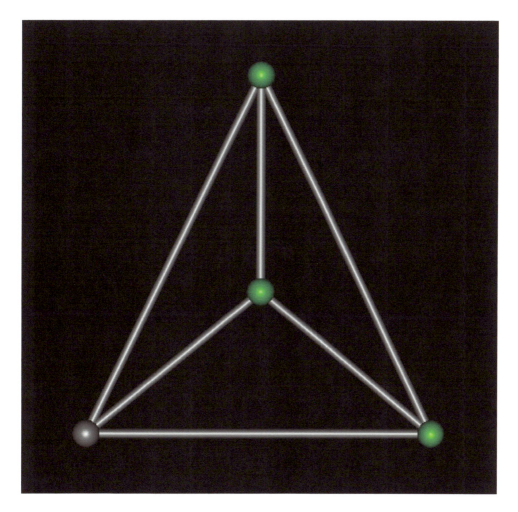

***Figure* 7.1.** *The graph of the Tetrahedron with a minimum vertex cover*
($n = 4$, $k = 3$).

7.2. The Kuratowski Bipartite Graph $K_{3,3}$ [9]. We run the program on the Kuratowski bipartite graph $K_{3,3}$ with $n = 6$ vertices. The algorithm finds a minimum vertex cover of size $k = 3$.

```
graph.txt
```

```
6
0 0 0 1 1 1
0 0 0 1 1 1
0 0 0 1 1 1
1 1 1 0 0 0
1 1 1 0 0 0
1 1 1 0 0 0
```

cover.txt
```
Vertex Cover (3): 4 5 6
```

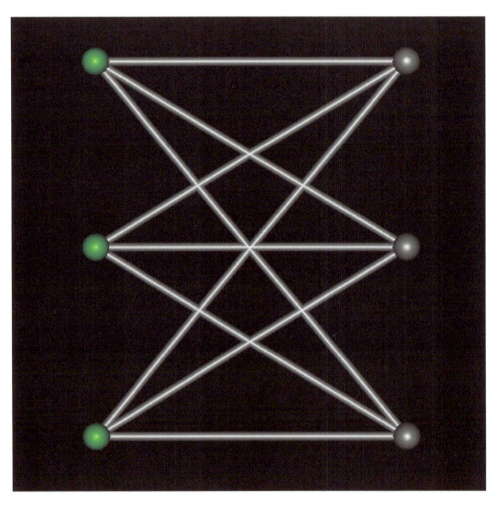

***Figure* 7.2.** *The Kuratowski graph $K_{3,3}$ with a minimum vertex cover*
($n = 6, k = 3$).

7.3. The Octahedron [8]. We run the program on the graph of the Octahedron with $n = 6$ vertices. The algorithm finds a minimum vertex cover of size $k = 4$.

graph.txt
```
6
0 1 1 0 1 1
1 0 1 1 0 1
1 1 0 1 1 0
```

```
0 1 1 0 1 1
1 0 1 1 0 1
1 1 0 1 1 0
```

cover.txt
```
Vertex Cover (4): 2 3 5 6
```

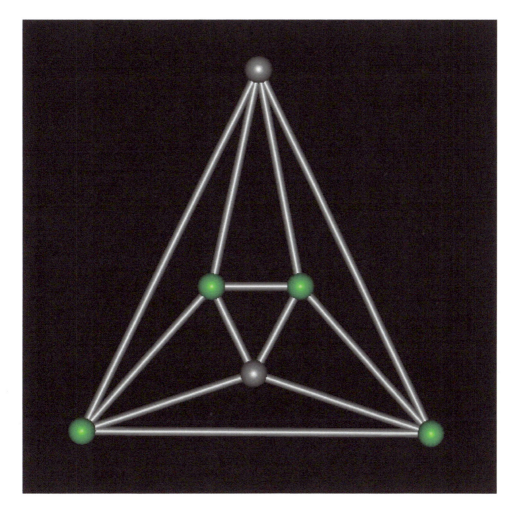

***Figure* 7.3.** *The graph of the Octahedron with a minimum vertex cover*
($n = 6$, $k = 4$).

7.4. The Bondy-Murty Graph G_1 [4]. We run the program on the Bondy-Murty graph G_1 with $n = 7$ vertices. The algorithm finds a minimum vertex cover of size $k = 4$.

graph.txt
```
7
0 1 1 0 1 1 0
1 0 1 1 0 1 0
1 1 0 1 1 0 0
0 1 1 0 0 0 1
1 0 1 0 0 0 1
1 1 0 0 0 0 1
0 0 0 1 1 1 0
```

```
cover.txt
Vertex Cover (4): 1 2 3 7
```

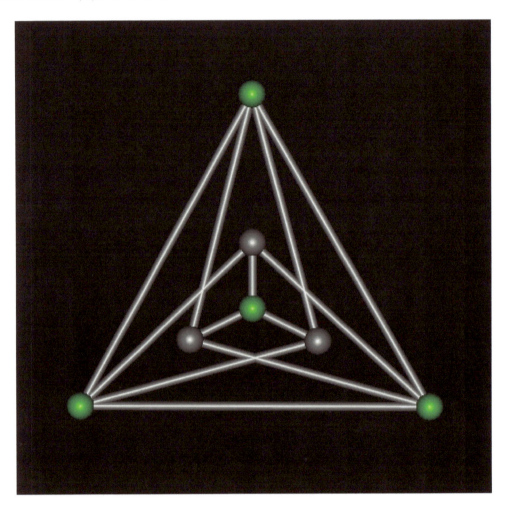

***Figure* 7.4.** *The Bondy-Murty graph G_1 with a minimum vertex cover*
($n = 7, k = 4$).

7.5. The Wheel Graph W_8 [4]. We run the program on the Wheel graph W_8 with $n = 8$ vertices. The algorithm finds a minimum vertex cover of size $k = 5$.

```
graph.txt
8
0 1 0 0 0 0 1 1
1 0 1 0 0 0 0 1
0 1 0 1 0 0 0 1
0 0 1 0 1 0 0 1
0 0 0 1 0 1 0 1
0 0 0 0 1 0 1 1
1 0 0 0 0 1 0 1
1 1 1 1 1 1 1 0
```

```
cover.txt
```

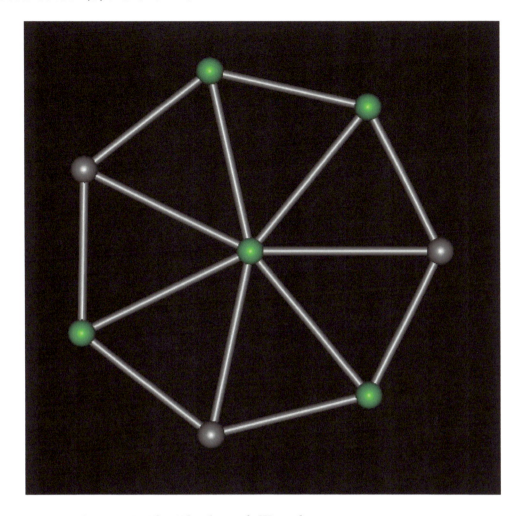

Figure 7.5. *The Wheel graph W_8 with a minimum vertex cover*
($n = 8$, $k = 5$).

7.6. The Cube [8]. We run the program on the graph of the Cube with $n = 8$ vertices. The algorithm finds a minimum vertex cover of size $k = 4$.

graph.txt
```
8
0 1 0 1 0 1 0 0
1 0 1 0 0 0 1 0
0 1 0 1 0 0 0 1
1 0 1 0 1 0 0 0
0 0 0 1 0 1 0 1
1 0 0 0 1 0 1 0
0 1 0 0 0 1 0 1
0 0 1 0 1 0 1 0
```

cover.txt
```
Vertex Cover (4):  2  4  6  8
```

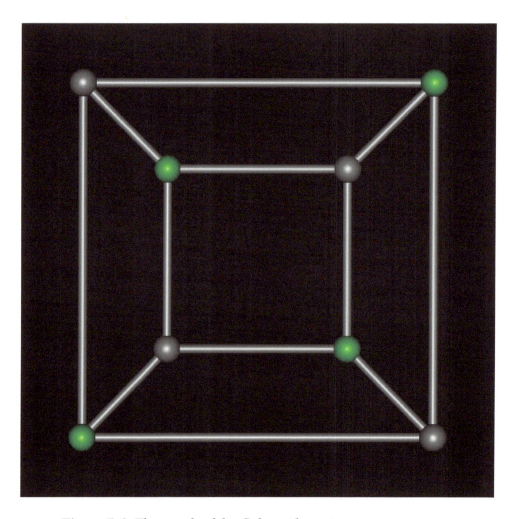

Figure 7.6. *The graph of the Cube with a minimum vertex cover*
(n = 8, k = 4).

7.7. The Petersen Graph [10]. We run the program on the Petersen graph with $n = 10$ vertices. The algorithm finds a minimum vertex cover of size $k = 6$.

```
graph.txt
10
0 1 0 0 1 0 1 0 0 0
1 0 1 0 0 0 0 1 0 0
0 1 0 1 0 0 0 0 1 0
0 0 1 0 1 0 0 0 0 1
1 0 0 1 0 1 0 0 0 0
0 0 0 0 1 0 0 1 1 0
1 0 0 0 0 0 0 0 1 1
0 1 0 0 0 1 0 0 0 1
0 0 1 0 0 1 1 0 0 0
0 0 0 1 0 0 1 1 0 0

cover.txt
Vertex Cover (6): 2 4 5 7 8 9
```

26

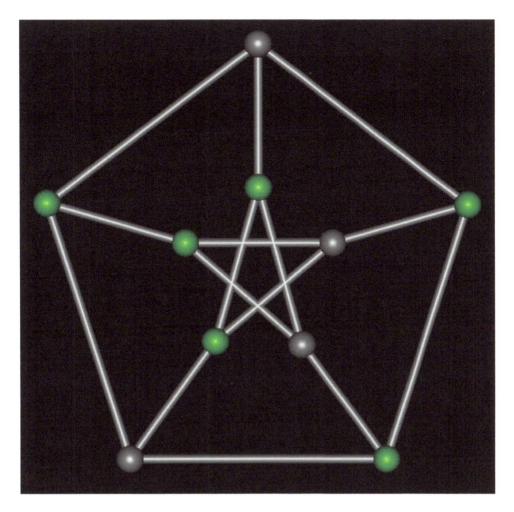

Figure 7.7. *The Petersen graph with a minimum vertex cover* ($n = 10$, $k = 6$).

7.8. The Bondy-Murty graph G_2 [4]. We run the program on the Bondy-Murty graph G_2 with $n = 11$ vertices. The algorithm finds a minimum vertex cover of size $k = 7$.

```
graph.txt
11
0 0 1 1 1 1 0 1 1 1 1
0 0 1 1 1 1 0 1 1 1 1
1 1 0 1 0 0 1 0 0 0 0
1 1 1 0 0 0 1 0 0 0 0
1 1 0 0 0 1 1 0 0 0 0
1 1 0 0 1 0 1 0 0 0 0
0 0 1 1 1 1 0 1 1 1 1
1 1 0 0 0 0 1 0 1 0 0
1 1 0 0 0 0 1 1 0 0 0
1 1 0 0 0 0 1 0 0 0 1
1 1 0 0 0 0 1 0 0 1 0

cover.txt
```

```
Vertex Cover (8):  3 4 5 6 8 9 10 11
Vertex Cover (8):  3 4 5 6 8 9 10 11
Vertex Cover (7):  1 2 4 5 7 9 10
```

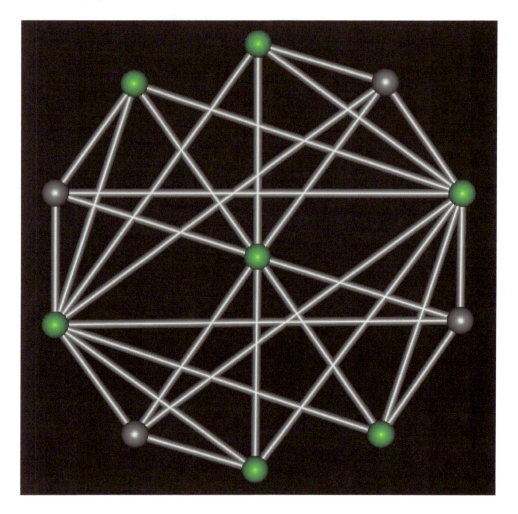

Figure *7.8. The Bondy-Murty graph* G_2 *with a minimum vertex cover*
($n = 11, k = 7$).

7.9. The Grötzsch Graph [11]. We run the program on the Grötzsch graph with $n = 11$
vertices. The algorithm finds a minimum vertex cover of size $k = 6$.

graph.txt
```
11
0 1 1 1 1 1 0 0 0 0 0
1 0 0 0 0 0 1 0 1 0 0
1 0 0 0 0 0 0 1 0 1 0
1 0 0 0 0 0 0 0 1 0 1
1 0 0 0 0 0 1 0 0 1 0
1 0 0 0 0 0 0 1 0 0 1
0 1 0 0 1 0 0 1 0 0 1
0 0 1 0 0 1 1 0 1 0 0
0 1 0 1 0 0 0 1 0 1 0
0 0 1 0 1 0 0 0 1 0 1
0 0 0 1 0 1 1 0 0 1 0
```

28

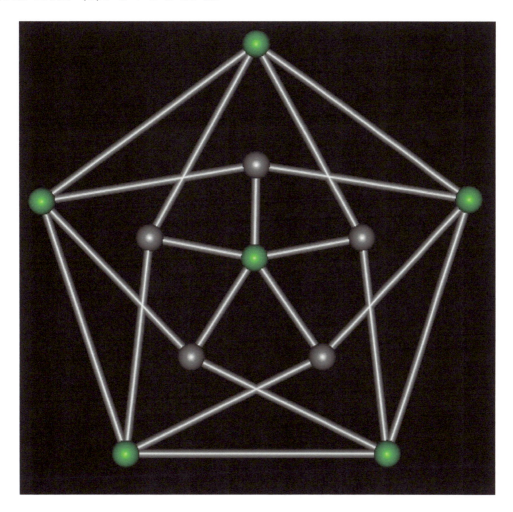

Figure 7.9. *The Grötzsch graph with a minimum vertex cover*
($n = 11, k = 6$).

7.10. The Herschel Graph [12]. We run the program on the Herschel graph with $n = 11$ vertices. The algorithm finds a minimum vertex cover of size $k = 5$.

graph.txt
```
11
0 1 0 1 1 0 1 0 0 0 0
1 0 1 0 0 0 0 1 0 0 0
0 1 0 1 0 0 0 0 1 0 0
1 0 1 0 0 0 0 0 0 1 0
1 0 0 0 0 1 0 0 0 1 0
0 0 0 0 1 0 1 0 0 0 1
1 0 0 0 0 1 0 1 0 0 0
0 1 0 0 0 0 1 0 1 0 1
0 0 1 0 0 0 0 1 0 1 0
```

```
0 0 0 1 1 0 0 0 1 0 1
0 0 0 0 0 1 0 1 0 1 0
```

cover.txt
```
Vertex Cover (6): 2 4 5 7 9 11
Vertex Cover (5): 1 3 6 8 10
```

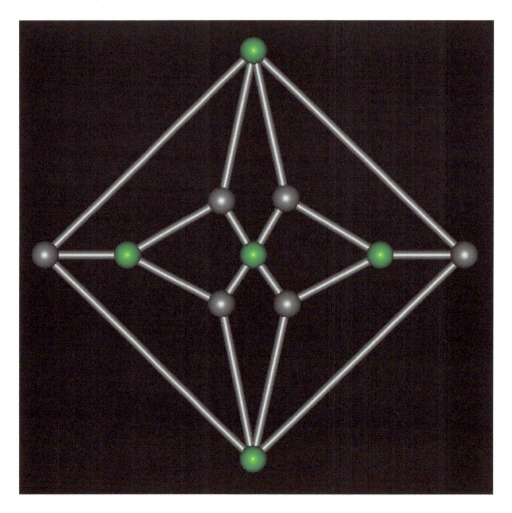

***Figure* 7.10.** *The Herschel graph with a minimum vertex cover*
($n = 11, k = 5$).

7.11. The Icosahedron [8]. We run the program on the graph of the Icosahedron with $n = 12$ vertices. The algorithm finds a minimum vertex cover of size $k = 9$.

graph.txt
```
12
 0 1 1 0 0 1 1 1 0 0 0 0
 1 0 1 1 1 1 0 0 0 0 0 0
 1 1 0 1 0 0 0 1 1 0 0 0
 0 1 1 0 1 0 0 0 1 1 0 0
 0 1 0 1 0 1 0 0 0 1 1 0
 1 1 0 0 1 0 1 0 0 0 1 0
 1 0 0 0 0 1 0 1 0 0 1 1
```

```
1 0 1 0 0 0 1 0 1 0 0 1
0 0 1 1 0 0 0 1 0 1 0 1
0 0 0 1 1 0 0 0 1 0 1 1
0 0 0 0 1 1 1 0 0 1 0 1
0 0 0 0 0 0 1 1 1 1 1 0
```

cover.txt
Vertex Cover (9): 1 2 3 5 6 7 9 10 12

***Figure* 7.11.** *The graph of the Icosahedron with a minimum vertex cover*
($n = 12, k = 9$).

7.12. The Bondy-Murty graph G_3 [4]. We run the program on the Bondy-Murty graph
G_3 with $n = 14$ vertices. The algorithm finds a minimum vertex cover of size $k = 7$.

graph.txt
```
14
0 0 0 1 0 0 0 0 1 0 0 0 1 0 0
0 0 1 0 0 0 1 0 0 0 1 0 0 0
0 1 0 1 0 0 0 0 0 0 0 0 0 1
1 0 1 0 1 0 0 0 0 0 0 0 0 0
0 0 0 1 0 1 0 1 0 0 0 1 0 0 0 0
```

```
0 0 0 0 1 0 1 0 0 0 0 0 1 0
0 1 0 0 0 1 0 1 0 0 0 0 0 0
1 0 0 0 0 0 1 0 1 0 0 0 0 0
0 0 0 0 0 0 0 1 0 1 0 0 0 1
0 0 0 0 1 0 0 0 1 0 1 0 0 0
0 1 0 0 0 0 0 0 0 1 0 1 0 0
1 0 0 0 0 0 0 0 0 0 1 0 1 0
0 0 0 0 0 1 0 0 0 0 0 1 0 1
0 0 1 0 0 0 0 0 1 0 0 0 1 0
```

cover.txt
```
Vertex Cover (7): 2 4 6 8 10 12 14
```

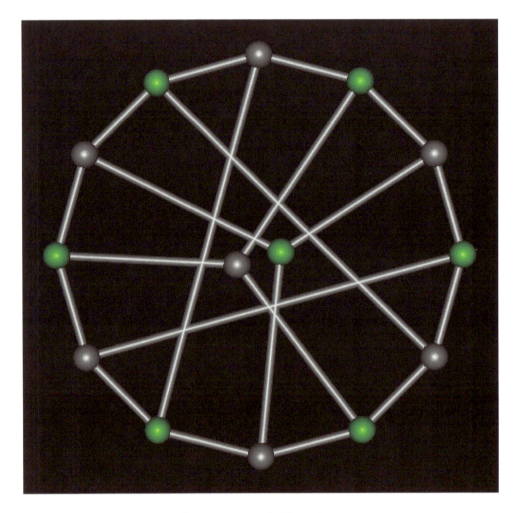

Figure 7.12. *The Bondy-Murty graph* G_3 *with a minimum vertex cover* ($n = 14, k = 7$).

7.13. The Bondy-Murty graph G_4 [4]. We run the program on the Bondy-Murty graph G_4 with $n = 16$ vertices. The algorithm finds a minimum vertex cover of size $k = 7$.

graph.txt
```
16
0 0 0 0 0 1 0 0 0 0 0 0 0 1 0 0 0
```

```
0 0 0 0 0 0 1 0 0 1 0 0 0 0 0 0
0 0 0 0 0 1 0 0 0 0 0 0 0 1 0 0 0
0 0 0 0 0 0 0 0 0 0 1 0 0 0 0 0
0 0 0 0 0 0 0 1 0 0 0 0 0 0 0 0
1 0 1 0 0 0 0 0 0 0 1 0 0 0 0 0
0 1 0 0 0 0 0 0 1 0 0 0 0 0 0 0
0 0 0 0 1 0 0 0 0 0 0 1 0 1 0 0
0 0 0 0 0 0 1 0 0 0 0 0 0 0 1 0
0 1 0 0 0 0 0 0 0 0 0 0 0 0 1 0
0 0 0 1 0 1 0 0 0 0 0 0 0 0 0 0
0 0 0 0 0 0 0 1 0 0 0 0 0 0 0 0
1 0 1 0 0 0 0 0 0 0 0 0 0 0 0 0
0 0 0 0 0 0 0 1 0 0 0 0 0 0 0 0
0 0 0 0 0 0 0 0 0 1 1 0 0 0 0 1
0 0 0 0 0 0 0 0 0 0 0 0 0 0 1 0
```

cover.txt
Vertex Cover (7): 2 6 7 8 11 13 15

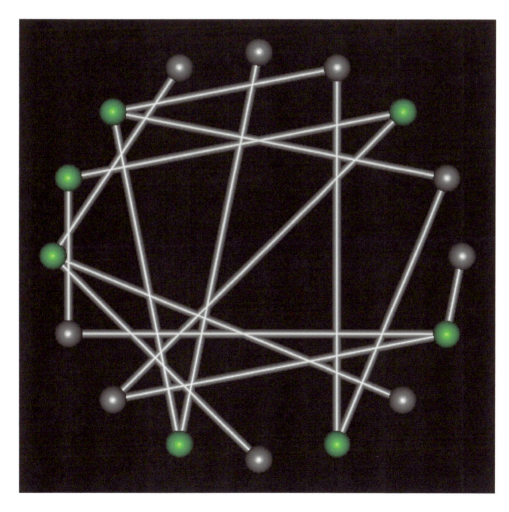

Figure 7.13. *The Bondy-Murty graph* G_4 *with a minimum vertex cover*
($n = 16, k = 7$).

7.14. The Ramsey Graph $R(4,4)$ [6]. We run the program on the Ramsey graph $R(4,4)$ with $n = 17$ vertices. The algorithm finds a minimum vertex cover of size $k = 14$.

graph.txt
```
17
0 1 1 0 1 0 0 0 1 1 0 0 0 1 0 1 1
1 0 1 1 0 1 0 0 0 1 1 0 0 0 1 0 1
1 1 0 1 1 0 1 0 0 0 1 1 0 0 0 1 0
0 1 1 0 1 1 0 1 0 0 0 1 1 0 0 0 1
1 0 1 1 0 1 1 0 1 0 0 0 1 1 0 0 0
0 1 0 1 1 0 1 1 0 1 0 0 0 1 1 0 0
0 0 1 0 1 1 0 1 1 0 1 0 0 0 1 1 0
0 0 0 1 0 1 1 0 1 1 0 1 0 0 0 1 1
1 0 0 0 1 0 1 1 0 1 1 0 1 0 0 0 1
1 1 0 0 0 1 0 1 1 0 1 1 0 1 0 0 0
0 1 1 0 0 0 1 0 1 1 0 1 1 0 1 0 0
0 0 1 1 0 0 0 1 0 1 1 0 1 1 0 1 0
0 0 0 1 1 0 0 0 1 0 1 1 0 1 1 0 1
1 0 0 0 1 1 0 0 0 1 0 1 1 0 1 1 0
0 1 0 0 0 1 1 0 0 0 1 0 1 1 0 1 1
1 0 1 0 0 0 1 1 0 0 0 1 0 1 1 0 1
1 1 0 1 0 0 0 1 1 0 0 0 1 0 1 1 0
```

cover.txt
```
Vertex Cover (14):  2 3 4 5 6 8 9 10 11 12 14 15 16 17
```

Figure 7.14. *The Ramsey graph R(4,4) with a minimum vertex cover*
($n = 17$, $k = 14$).

7.15. The Folkman Graph [13]. We run the program on the Folkman graph with $n = 20$ vertices. The algorithm finds a minimum vertex cover of size $k = 10$.

graph.txt
```
20
0 0 0 0 0 0 0 0 0 0 1 0 1 0 0 0 0 0 1 1
0 0 0 0 0 0 0 0 0 0 0 1 0 1 0 0 0 1 1 0
0 0 0 0 0 0 0 0 0 0 0 0 1 0 1 0 1 1 0 0
0 0 0 0 0 0 0 0 0 0 1 0 0 1 0 1 1 0 0 0
0 0 0 0 0 0 0 0 0 0 0 1 0 0 1 1 0 0 0 1
0 0 0 0 0 0 0 0 0 0 0 1 0 0 1 1 0 0 0 1
0 0 0 0 0 0 0 0 0 0 1 0 0 1 0 1 1 0 0 0
0 0 0 0 0 0 0 0 0 0 0 1 0 1 0 1 1 0 0 0
0 0 0 0 0 0 0 0 0 0 0 1 0 1 0 0 0 1 1 0
0 0 0 0 0 0 0 0 0 0 1 0 1 0 0 0 0 0 1 1
1 0 0 1 0 0 1 0 0 1 0 0 0 0 0 0 0 0 0 0
0 1 0 0 1 1 0 0 1 0 0 0 0 0 0 0 0 0 0 0
1 0 1 0 0 0 0 1 0 1 0 0 0 0 0 0 0 0 0 0
0 1 0 1 0 0 1 0 1 0 0 0 0 0 0 0 0 0 0 0
0 0 1 0 1 1 0 1 0 1 0 0 0 0 0 0 0 0 0 0
```

```
0 0 0 1 1 1 1 0 0 0 0 0 0 0 0 0 0 0 0 0
0 0 1 1 0 0 1 1 0 0 0 0 0 0 0 0 0 0 0 0
0 1 1 0 0 0 0 1 1 0 0 0 0 0 0 0 0 0 0 0
1 1 0 0 0 0 0 0 1 1 0 0 0 0 0 0 0 0 0 0
1 0 0 0 1 1 0 0 0 1 0 0 0 0 0 0 0 0 0 0
```

cover.txt
```
Vertex Cover (10): 11 12 13 14 15 16 17 18 19 20
```

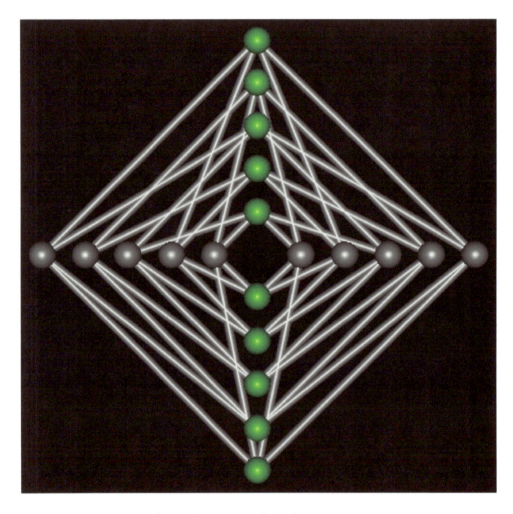

Figure 7.15. *The Folkman graph with a minimum vertex cover*
($n = 20, k = 10$).

7.16. The Dodecahedron [8]. We run the program on the graph of the Dodecahedron with $n = 20$ vertices. The algorithm finds a minimum vertex cover of size $k = 12$.

graph.txt
```
20
0 1 0 0 1 0 0 0 0 0 0 0 0 0 1 0 0 0 0 0
1 0 1 0 0 0 0 0 0 0 0 1 0 0 0 0 0 0 0 0
0 1 0 1 0 0 0 0 0 1 0 0 0 0 0 0 0 0 0 0
0 0 1 0 1 0 0 1 0 0 0 0 0 0 0 0 0 0 0 0
1 0 0 1 0 1 0 0 0 0 0 0 0 0 0 0 0 0 0 0
```

```
0 0 0 0 1 0 1 0 0 0 0 0 0 0 0 1 0 0 0 0 0
0 0 0 0 0 1 0 1 0 0 0 0 0 0 0 0 1 0 0 0
0 0 0 1 0 0 1 0 1 0 0 0 0 0 0 0 0 0 0 0
0 0 0 0 0 0 0 1 0 1 0 0 0 0 0 0 0 1 0 0
0 0 1 0 0 0 0 0 1 0 1 0 0 0 0 0 0 0 0 0
0 0 0 0 0 0 0 0 0 1 0 1 0 0 0 0 0 0 1 0
0 1 0 0 0 0 0 0 0 0 1 0 1 0 0 0 0 0 0 0
0 0 0 0 0 0 0 0 0 0 0 1 0 1 0 0 0 0 0 1
1 0 0 0 0 0 0 0 0 0 0 0 1 0 1 0 0 0 0 0
0 0 0 0 0 1 0 0 0 0 0 0 0 1 0 1 0 0 0 0
0 0 0 0 0 0 0 0 0 0 0 0 0 0 1 0 1 0 0 1
0 0 0 0 0 0 1 0 0 0 0 0 0 0 0 1 0 1 0 0
0 0 0 0 0 0 0 0 1 0 0 0 0 0 0 0 1 0 1 0
0 0 0 0 0 0 0 0 0 0 1 0 0 0 0 0 0 1 0 1
0 0 0 0 0 0 0 0 0 0 0 0 1 0 0 1 0 0 1 0
```

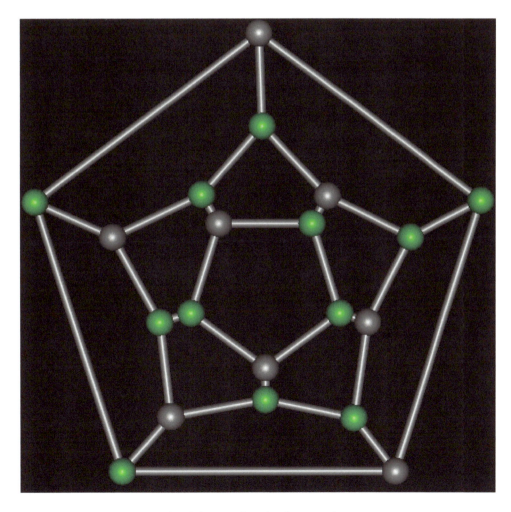

Figure 7.16. *The graph of the Dodecahedron with a minimum vertex cover* (*n* = 20, *k* = 12).

7.17. The Tutte-Coxeter Graph [14]. We run the program on the Tutte-Coxeter graph with $n = 30$ vertices. The algorithm finds a minimum vertex cover of size $k = 15$.

graph.txt
```
30
0 1 0 0 0 0 0 0 0 0 0 0 0 0 0 0 0 0 0 0 0 1 0 0 0 0 0 0 0 1
1 0 1 0 0 0 0 0 1 0 0 0 0 0 0 0 0 0 0 0 0 0 0 0 0 0 0 0 0 0
0 1 0 1 0 0 0 0 0 0 0 0 0 0 0 0 0 0 0 0 0 0 0 0 1 0 0 0 0 0
0 0 1 0 1 0 0 0 0 0 0 0 1 0 0 0 0 0 0 0 0 0 0 0 0 0 0 0 0 0
0 0 0 1 0 1 0 0 0 0 0 0 0 0 0 0 1 0 0 0 0 0 0 0 0 0 0 0 0 0
0 0 0 0 1 0 1 0 0 0 0 0 0 0 0 0 0 0 0 0 0 1 0 0 0 0 0 0 0 0
0 0 0 0 0 1 0 1 0 0 0 0 0 0 0 0 0 0 0 0 0 0 0 0 0 1 0 0
0 0 0 0 0 0 1 0 1 0 0 0 0 0 1 0 0 0 0 0 0 0 0 0 0 0 0 0 0 0
0 1 0 0 0 0 0 1 0 1 0 0 0 0 0 0 0 0 0 0 0 0 0 0 0 0 0 0 0 0
0 0 0 0 0 0 0 0 1 0 1 0 0 0 0 0 0 0 1 0 0 0 0 0 0 0 0 0 0 0
0 0 0 0 0 0 0 0 0 1 0 1 0 0 0 0 0 0 0 0 0 0 0 1 0 0 0 0 0 0
0 0 0 0 0 0 0 0 0 0 1 0 1 0 0 0 0 0 0 0 0 0 0 0 0 0 0 0 1 0
0 0 0 1 0 0 0 0 0 0 0 1 0 1 0 0 0 0 0 0 0 0 0 0 0 0 0 0 0 0
0 0 0 0 0 0 0 0 0 0 0 0 1 0 1 0 0 0 0 0 1 0 0 0 0 0 0 0 0 0
0 0 0 0 0 0 0 1 0 0 0 0 0 1 0 1 0 0 0 0 0 0 0 0 0 0 0 0 0 0
0 0 0 0 0 0 0 0 0 0 0 0 0 0 1 0 1 0 0 0 0 0 0 1 0 0 0 0 0 0
0 0 0 0 0 0 0 0 0 0 0 0 0 0 0 1 0 1 0 0 0 0 0 0 0 0 0 0 0 1
0 0 0 0 1 0 0 0 0 0 0 0 0 0 0 0 1 0 1 0 0 0 0 0 0 0 0 0 0 0
0 0 0 0 0 0 0 0 1 0 0 0 0 0 0 0 0 1 0 1 0 0 0 0 0 0 0 0 0 0
0 0 0 0 0 0 0 0 0 0 0 0 0 0 0 0 0 0 1 0 1 0 0 0 0 0 1 0 0 0
0 0 0 0 0 0 0 0 0 0 0 0 0 1 0 0 0 0 0 1 0 1 0 0 0 0 0 0 0 0
1 0 0 0 0 0 0 0 0 0 0 0 0 0 0 0 0 0 0 0 1 0 1 0 0 0 0 0 0 0
0 0 0 0 0 1 0 0 0 0 0 0 0 0 0 0 0 0 0 0 0 1 0 1 0 0 0 0 0 0
0 0 0 0 0 0 0 0 0 0 1 0 0 0 0 0 0 0 0 0 0 0 1 0 1 0 0 0 0 0
0 0 0 0 0 0 0 0 0 0 0 0 0 0 0 1 0 0 0 0 0 0 0 1 0 1 0 0 0 0
0 0 1 0 0 0 0 0 0 0 0 0 0 0 0 0 0 0 0 0 0 0 0 0 1 0 1 0 0 0
0 0 0 0 0 0 0 0 0 0 0 0 0 0 0 0 0 1 0 0 0 0 0 0 0 1 0 1 0 0
0 0 0 0 0 0 1 0 0 0 0 0 0 0 0 0 0 0 0 0 0 0 0 0 0 0 1 0 1 0
0 0 0 0 0 0 0 0 0 0 0 1 0 0 0 0 0 0 0 0 0 0 0 0 0 0 0 1 0 1
1 0 0 0 0 0 0 0 0 0 0 0 0 0 0 1 0 0 0 0 0 0 0 0 0 0 0 0 1 0
```

cover.txt
```
Vertex Cover (15): 1 3 5 7 9 11 13 15 17 19 21 23 25 27 29
```

38

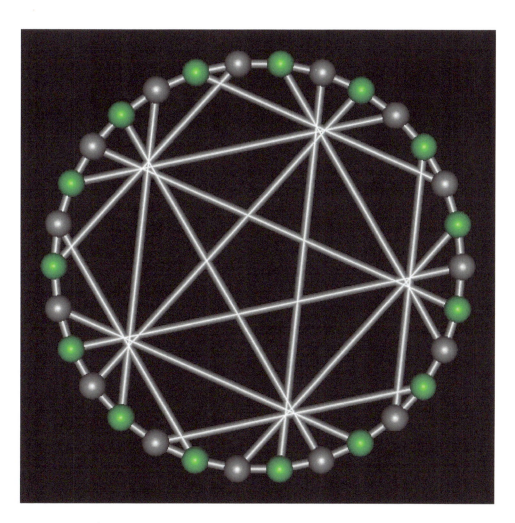

Figure 7.17. *The Tutte-Coxeter graph with a minimum vertex cover*
(*n* = 30, *k* = 15).

7.18. The Thomassen Graph [15]. We run the program on the Thomassen graph with n = 34 vertices. The algorithm finds a minimum vertex cover of size $k = 20$.

graph.txt

```
34
0 0 1 1 0 1 0 0 0 0 0 0 0 0 0 0 0 0 0 0 0 0 0 0 0 0 0 0 0 0 0 0 0 0
0 0 0 1 1 0 0 0 0 0 0 0 0 1 0 0 0 0 0 0 0 0 0 0 0 0 0 0 0 0 0 0 0 0
1 0 0 0 1 0 0 0 0 0 0 0 0 0 1 0 0 0 0 0 0 0 0 0 0 0 0 0 0 0 0 0 0 0
1 1 0 0 0 0 0 0 0 0 0 0 0 0 0 1 0 0 0 0 0 0 0 0 0 0 0 0 0 0 0 0 0 0
0 1 1 0 0 0 0 0 0 0 0 0 0 0 0 0 1 0 0 0 0 0 0 0 0 0 0 0 0 0 0 0 0 0
1 0 0 0 0 0 0 1 1 0 0 0 0 0 0 0 0 0 0 0 0 0 0 0 0 0 0 0 0 0 0 0 0 0
0 0 0 0 0 0 0 0 1 1 0 0 1 0 0 0 0 0 0 0 0 0 0 0 0 0 0 0 0 0 0 0 0 0
0 0 0 0 0 1 0 0 0 1 0 1 0 0 0 0 0 0 0 0 0 0 0 0 0 0 0 0 0 0 0 0 0 0
0 0 0 0 0 1 1 0 0 0 1 0 0 0 0 0 0 0 0 0 0 0 0 0 0 0 0 0 0 0 0 0 0 0
0 0 0 0 0 0 1 1 0 0 0 0 0 0 0 0 0 1 0 0 0 0 0 0 0 0 0 0 0 0 0 0 0 0
0 0 0 0 0 0 0 0 1 0 0 1 0 0 0 0 0 0 0 1 0 0 0 0 0 0 0 0 0 0 0 0 0 0
0 0 0 0 0 0 0 1 0 0 1 0 1 0 0 0 0 0 0 0 0 0 0 0 0 0 0 0 0 0 0 0 0 0
0 0 0 0 0 0 1 0 0 0 1 0 1 0 0 0 0 0 0 0 0 0 0 0 0 0 0 0 0 0 0 0 0 0
0 1 0 0 0 0 0 0 0 0 0 0 1 0 1 0 0 0 0 0 0 0 0 0 0 0 0 0 0 0 0 0 0 0
0 0 1 0 0 0 0 0 0 0 0 0 0 1 0 1 0 0 0 0 0 0 0 0 0 0 0 0 0 0 0 0 0 0
0 0 0 1 0 0 0 0 0 0 0 0 0 0 1 0 1 0 0 0 0 0 0 0 0 0 0 0 0 0 0 0 0 0
0 0 0 0 1 0 0 0 0 0 0 0 0 0 0 1 0 1 0 0 0 0 0 1 0 0 0 0 0 0 0 0 0 0
0 0 0 0 0 0 0 0 0 1 0 0 0 0 0 0 1 0 1 0 0 0 0 0 1 0 0 0 0 0 0 0 0 0
0 0 0 0 0 0 0 0 0 0 0 0 0 0 0 0 0 1 0 1 0 0 0 0 0 1 0 0 0 0 0 0 0 0
0 0 0 0 0 0 0 0 0 0 1 0 0 0 0 0 0 0 1 0 1 0 0 0 0 0 1 0 0 0 0 0 0 0
0 0 0 0 0 0 0 0 0 0 0 0 0 0 0 0 0 0 0 1 0 1 0 0 0 0 0 0 0 0 0 0 0 1
0 0 0 0 0 0 0 0 0 0 0 0 0 0 0 0 0 0 0 0 1 0 1 0 0 0 0 0 0 0 0 0 1 0
0 0 0 0 0 0 0 0 0 0 0 0 0 0 0 0 0 0 0 0 0 1 0 1 0 0 0 0 0 0 1 0 0 0
0 0 0 0 0 0 0 0 1 1 0 0 0 0 0 0 0 0 0 0 0 0 1 0 0 0 0 0 1 0 0 0 0 0
0 0 0 0 0 0 0 0 0 0 0 0 0 0 0 0 0 0 0 0 0 0 1 0 0 1 1 0 1 0 0 0 0 0
0 0 0 0 0 0 0 0 0 0 0 0 0 0 0 0 0 1 0 0 0 0 0 0 0 0 1 1 0 0 0 0 0 0
0 0 0 0 0 0 0 0 0 0 0 0 0 0 0 0 0 0 1 0 0 0 0 0 1 0 0 1 0 0 0 0 0 0
0 0 0 0 0 0 0 0 0 0 0 0 0 0 0 0 0 0 0 1 0 0 0 0 0 1 1 0 0 0 0 0 0 0
0 0 0 0 0 0 0 0 0 0 0 0 0 0 0 0 0 0 0 0 1 0 0 0 0 0 1 1 0 0 0 0 0 0
0 0 0 0 0 0 0 0 0 0 0 0 0 0 0 0 0 0 0 0 0 0 1 0 0 0 0 0 1 1 0 0 0 0
0 0 0 0 0 0 0 0 0 0 0 0 0 0 0 0 0 0 0 0 0 0 0 1 0 0 0 0 0 0 0 1 1 1
0 0 0 0 0 0 0 0 0 0 0 0 0 0 0 0 0 0 0 0 0 1 0 0 0 0 0 0 1 0 0 0 1 0
0 0 0 0 0 0 0 0 0 0 0 0 0 0 0 0 0 0 0 0 0 0 1 0 0 0 0 0 0 1 1 0 0 0
0 0 0 0 0 0 0 0 0 0 0 0 0 0 0 0 0 0 0 0 0 0 0 1 0 0 0 0 0 0 1 1 0 0
```

cover.txt

```
Vertex Cover (20): 1 2 5 8 9 10 11 13 15 16 18 20 22 23 26 27 28 30 31
34
```

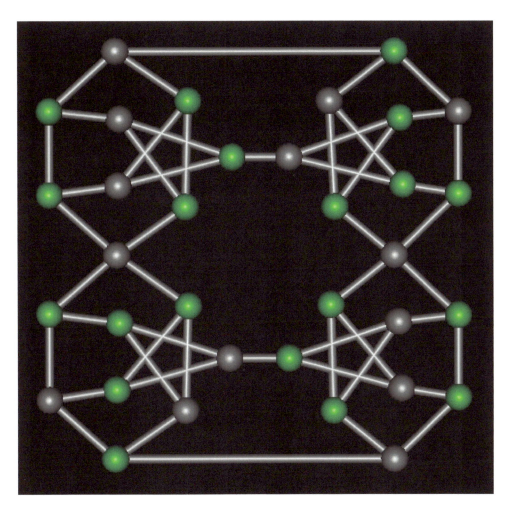

Figure 7.18. *The Thomassen graph with a minimum vertex cover*
(*n* = 34, *k* = 20).

7.19. The Berge Graph [16]. This is the first benchmark graph with $n = 60$ vertices, following a construction due to Claude Berge. Let G denote the graph of the Dodecahedron and let $H = K_3$ denote the graph of the Triangle i.e. the clique on three vertices. The *Berge graph $G \times H$* is defined as the graph whose set of vertices is $V(G) \times V(H)$ with an edge connecting vertex (u_1, v_1) with vertex (u_2, v_2) if and only if either $u_1 = u_2$ and $\{v_1, v_2\}$ is an edge in H or $v_1 = v_2$ and $\{u_1, u_2\}$ is an edge in G. It is known that the vertices of the Dodecahedron can be properly coloured with three colours. As a consequence, the Berge graph should have a vertex cover with at most forty vertices. Indeed, the algorithm finds a minimum vertex cover of size $k = 40$.

graph.txt
[download]

cover.txt
Vertex Cover (3): 1 3 4 5 8 9 10 12 13 14 17 18 19 21 23 24 25 27 28 29
31 33 35 36 37 39 41 42 43 44 46 48 49 50 53 54 55 56 59 60

41

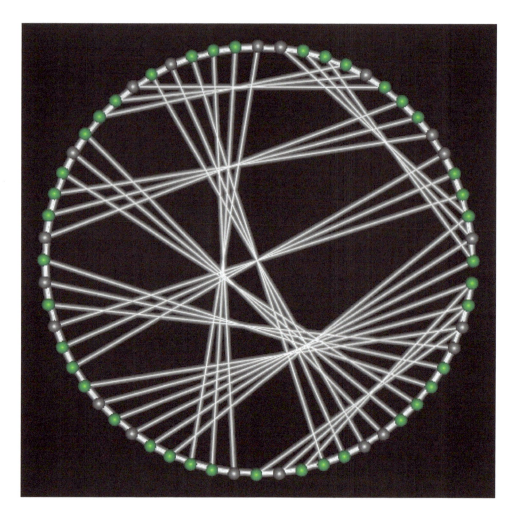

Figure 7.19. *The Berge graph with a minimum vertex cover*
(*n* = 60, *k* = 40).

7.20. The Witzel Graph [17]. This is the second benchmark graph with $n = 450$ vertices, following a construction due to Klaus D. Witzel. Take thirty disjoint cliques on fifteen vertices and connect random pairs of cliques by random edges. Shuffle the labels of the vertices well so that the original cliques are hidden. Provided this is done carefully without adding too many extra edges, such a graph should have a minimum vertex cover with at most 420 vertices (all but one vertex from each original clique). Moreover, the minimum vertex cover is well and truly hidden. Indeed, the algorithm finds a minimum vertex cover of size $k = 420$.

graph.txt
[download]

cover.txt
```
Vertex Cover (420):  1 2 3 4 6 7 8 9 10 11 12 13 14 15 16 17 18 20 21 22
23 24 25 26 27 28 29 30 31 32 33 35 36 37 38 39 40 41 42 43 44 45 46 47
48 49 50 51 52 53 54 56 57 58 59 60 61 62 63 64 65 66 67 68 69 70 71 72
73 75 76 77 79 80 81 82 83 84 85 86 87 88 89 90 91 92 93 94 95 96 98 99
100 101 102 103 104 105 106 107 108 109 110 111 112 113 114 115 116 117
```

42

```
118 119 121 123 124 125 126 127 128 129 130 131 132 133 134 135 136 137
138 139 140 141 143 144 145 146 147 148 149 150 151 152 153 154 155 156
157 158 160 161 162 163 164 165 166 168 169 170 171 172 173 174 175 176
177 178 179 180 181 182 183 184 185 187 188 189 190 191 192 193 194 195
196 197 198 199 200 201 202 203 204 205 207 208 209 210 212 213 214 215
216 217 218 219 220 221 222 223 224 225 226 228 229 230 231 232 233 234
235 236 237 238 239 240 241 243 244 245 246 247 248 249 250 251 252 253
254 255 256 257 258 259 260 261 262 263 264 265 266 267 269 270 271 272
273 274 275 276 277 278 279 281 282 283 284 285 286 287 288 289 290 291
292 293 294 295 296 297 299 300 301 302 303 304 305 306 307 308 309 310
311 312 313 315 316 317 318 319 320 321 322 323 324 325 326 327 328 330
331 332 333 334 335 337 338 339 340 341 342 343 344 345 346 347 348 349
350 352 353 354 355 356 357 358 359 360 361 362 363 365 366 367 368 369
370 371 372 373 374 375 376 377 378 379 380 381 382 383 385 386 387 388
389 390 391 392 393 394 395 396 397 398 399 401 402 403 404 405 406 407
408 409 410 412 413 414 415 416 417 418 419 420 421 422 423 424 425 427
428 429 430 431 432 433 434 435 436 437 438 439 440 441 443 444 445 446
447 448 449 450
```

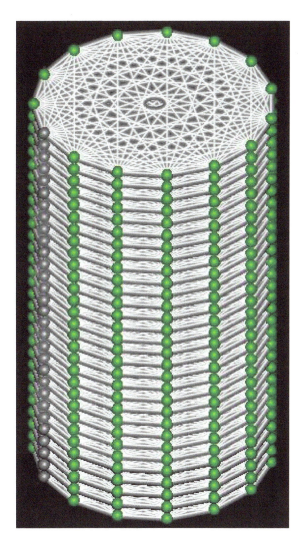

Figure 7.20. *The Witzel graph (scheme only) with a minimum vertex cover*
(n = 450, k = 420).

43

8. References

[1]   R.M. Karp, *Reducibility among combinatorial problems*, Complexity of Computer Computations, Plenum Press, 1972.

[2]   R. Frucht, *Graphs of degree three with a given abstract group*, Canad. J. Math., 1949.

[3]   Stephen Cook, *The **P** versus **NP** Problem*, Official Problem Description, Millennium Problems, Clay Mathematics Institute, 2000.

[4]   J.A. Bondy and U.S.R. Murty, *Graph Theory with Applications*, Elsevier Science Publishing Co., Inc, 1976.

[5]   Euclid, *Elements*, circa 300 B.C.

[6]   F.P. Ramsey, *On a problem of formal logic*, Proc. London Math. Soc., 1930.

[7]   Stanley Lippman, *Essential C++*, Addison-Wesley, 2000.

[8]   Plato, *Timaeaus*, circa 350 B.C.

[9]   K. Kuratowski, *Sur le problème des courbes gauches en topologie*, Fund. Math., 1930.

[10] J. Petersen, *Die Theorie der regulären Graphen*, Acta Math., 1891.

[11] H. Grötzsch, *Ein Dreifarbensatz für dreikreisfreie Netz auf der Kugel*, Z. Martin-Luther-Univ., 1958.

[12] A.S. Herschel, *Sir Wm. Hamilton's Icosian Game*, Quart. J. Pure Applied Math., 1862.

[13] J. Folkman, *Regular line-symmetric graphs*, J. Combinatorial Theory, 1967.

[14] H.S.M. Coxeter and W.T. Tutte, *The Chords of the Non-Ruled Quadratic in PG(3,3)*, Canad. J. Math., 1958.

[15] C. Thomassen, *Hypohamiltonian and hypotraceable graphs*, Discrete Math., 1974.

[16] C. Berge, *Graphes et Hypergraphes*, Dunod, 1970.

[17] Klaus D. Witzel, *Personal Communication*, 2006.

[18] Ashay Dharwadker, *The Independent Set Algorithm*, **http://www.dharwadker.org/independent_set** , 2006.

[19] Ashay Dharwadker, *The Clique Algorithm*, **http://www.dharwadker.org/clique** , 2006.

[20] Ashay Dharwadker, *The Vertex Coloring Algorithm*, **http://www.dharwadker.org/vertex_coloring** , 2006.